Table of Contents

- **Introduction**
- **Section 1: Easy Science Facts**

1. Biology
- Fun Animal Facts
- Amazing Plant Facts
- Human Body Wonders

2. Chemistry
- Elements and Compounds
- Interesting Chemical Reactions
- Everyday Chemistry

3. Physics
- Basic Principles of Physics
- Light, Sound, and Energy
- Everyday Physics Examples

4. Earth Science
- Weather, Rocks, and Minerals
- Volcanoes, Earthquakes, and •Natural Phenomena

- The Water Cycle and Climate Change

• Section 2: History Facts

1. Ancient Civilizations

- Egyptian Pyramids and Mummies
- Greek Mythology and the Olympics
- Roman Empire and Their Architecture

2. Medieval Times

- Knights and Castles
- Famous Kings and Queens
- Life in the Medieval Period

3. Modern History

- Important Inventions and Discoveries
- Notable Historical Figures
- Major Events

5. Indian History

- Ancient India
- Medieval India
- Mughal India

- Modern India

● Section 3: General Knowledge Facts

1. Human Body

- Amazing Facts About Our Organs
- Our Senses
- Fun Human Capabilities

2. Inventions and Discoveries

- Groundbreaking Inventions
- Famous Inventors
- Surprising Discoveries

3. World Records

- Tallest, Fastest, Smallest, and •Largest Records
- Extraordinary Human Feats
- Amazing Records in Nature and Technology

4. Language and Literature

- Interesting Facts About Languages
- Famous Authors and Their Works
- Unique Literary Traditions

●Section 4: Geography Facts

1.Continents and Countries

- Facts About Each Continent
- Unique Aspects of Different Countries
- Cultural Diversity Around the World

2.Mountains, Rivers, and Deserts

- Tallest Mountains and Longest Rivers
- Largest Deserts and Their Features
- Major Natural Landmarks

3.Oceans and Seas

- Interesting Marine Life
- Major Oceans and Seas
- Facts About Underwater Exploration

4.Maps and Globes

- How to Read Maps and Globes
- Important Lines Like the Equator and Prime Meridian
- Fascinating Cartography Facts

● Section 5: Space Facts

1. Solar System

- Planets, Moons, and the Sun
- Interesting Facts About Each Planet
- Asteroids, Comets, and Meteoroids

2. Stars and Galaxies

- Different Types of Stars
- Facts About the Milky Way and Other Galaxies
- The Life Cycle of Stars

3. Space Exploration

- Famous Space Missions and Astronauts
- Interesting Facts About Space Travel
- The International Space Station and Future Missions

4. Universe Mysteries

- Black Holes and Dark Matter
- Theories About the Origin of the Universe
- Fun Trivia About Space Phenomena

Introduction

Welcome to "GK ESSENTIALS"! This book is packed with fascinating facts and trivia that will take you on a journey through science, history, geography, space, and much more. Whether you're curious about the tallest mountains, the brightest stars, or the most interesting animals, you'll find plenty of exciting information here.

General Knowledge (GK) is all about knowing a little bit of everything. It helps you understand the world around you and makes learning fun. So, let's dive in and discover the wonders of our amazing world together!

BIOLOGY
1. FUN ANIMAL FACTS

ELEPHANTS ARE THE ONLY ANIMALS THAT CAN'T JUMP. THEY ARE THE LARGEST LAND ANIMALS AND CAN WEIGH UP TO 14,000 POUNDS!

DOLPHINS SLEEP WITH ONE EYE OPEN. THEY NEED TO BE ALERT TO POTENTIAL DANGERS EVEN WHILE RESTING.

HONEYBEES CAN RECOGNIZE HUMAN FACES. THEY CAN ALSO FLY UP TO 15 MILES PER HOUR AND VISIT 50 TO 100 FLOWERS DURING ONE COLLECTION TRIP.

2. AMAZING PLANT FACTS

BAMBOO CAN GROW UP TO 35 INCHES IN A SINGLE DAY. IT'S ONE OF THE FASTEST-GROWING PLANTS IN THE WORLD.

SUNFLOWERS CAN HELP CLEAN RADIOACTIVE SOIL. THEY WERE USED TO REMOVE TOXINS AFTER THE CHERNOBYL NUCLEAR DISASTER.

VENUS FLYTRAPS ARE CARNIVOROUS PLANTS THAT EAT INSECTS. THEY SNAP SHUT WHEN THEIR TINY HAIRS ARE TRIGGERED.

3. HUMAN BODY WONDERS

THE HEART BEATS ABOUT 100,000 TIMES A DAY, PUMPING BLOOD THROUGH 60,000 MILES OF BLOOD VESSELS.

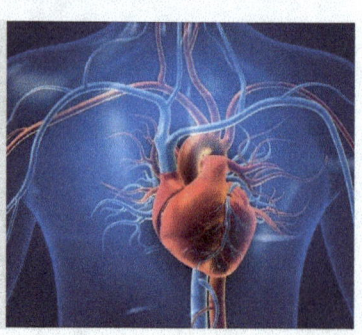

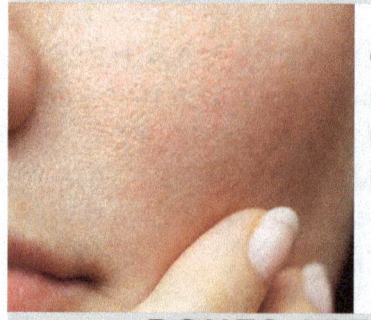

OUR SKIN IS THE BODY'S LARGEST ORGAN AND CAN RENEW ITSELF EVERY 28 DAYS.

BONES ARE FIVE TIMES STRONGER THAN STEEL OF THE SAME DENSITY. THE HUMAN BODY HAS 206 BONES IN TOTAL.

CHEMISTRY
1. ELEMENTS AND COMPOUNDS

WATER (H2O) IS MADE OF TWO HYDROGEN ATOMS AND ONE OXYGEN ATOM. IT'S ESSENTIAL FOR ALL KNOWN FORMS OF LIFE.

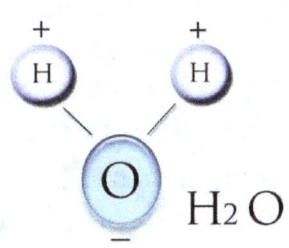

GOLD (AU) IS A VERY HEAVY METAL. A CUBIC FOOT OF GOLD WEIGHS AROUND 1,200 POUNDS.

TABLE SALT (NACL) IS MADE UP OF SODIUM AND CHLORINE. SODIUM IS A METAL, AND CHLORINE IS A GAS, BUT TOGETHER THEY FORM A COMMON SEASONING.

2. INTERESTING CHEMICAL REACTIONS

BAKING SODA AND VINEGAR CREATE A FIZZY REACTION. THIS IS OFTEN USED IN HOMEMADE VOLCANO PROJECTS.

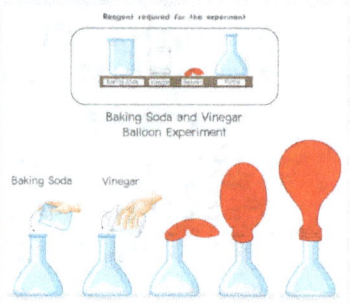

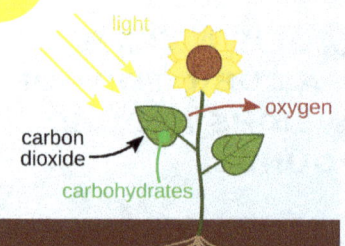

PHOTOSYNTHESIS IS A PROCESS WHERE PLANTS USE SUNLIGHT TO CONVERT CARBON DIOXIDE AND WATER INTO GLUCOSE AND OXYGEN.

RUSTING OCCURS WHEN IRON REACTS WITH OXYGEN AND WATER. THIS CHEMICAL REACTION FORMS IRON OXIDE, COMMONLY KNOWN AS RUST.

3. EVERYDAY CHEMISTRY

COOKING AN EGG CHANGES ITS STRUCTURE. THE HEAT CAUSES THE PROTEINS TO DENATURE AND COAGULATE, TURNING THE EGG FROM LIQUID TO SOLID.

SOAP WORKS BY BREAKING DOWN GREASE AND DIRT. IT HAS MOLECULES THAT ATTACH TO BOTH WATER AND OILS, MAKING IT EASY TO WASH AWAY GRIME.

YEAST IN BREAD PRODUCES CARBON DIOXIDE GAS, WHICH MAKES THE DOUGH RISE. THIS PROCESS IS CALLED FERMENTATION.

PHYSICS
1. BASIC PRINCIPLES OF PHYSICS

GRAVITY IS THE FORCE THAT PULLS OBJECTS TOWARDS EACH OTHER. IT'S WHY WE STAY ON THE GROUND AND WHY APPLES FALL FROM TREES.

Gravity pulls everything downwards

An object at rest will remain at rest unless a net force acts on it.

INERTIA IS THE TENDENCY OF AN OBJECT TO RESIST CHANGES IN ITS STATE OF MOTION. A ROLLING BALL WILL KEEP MOVING UNTIL SOMETHING STOPS IT.

An object in motion will remain in motion, unless a net force acts on it.

ENERGY CAN NEITHER BE CREATED NOR DESTROYED. IT ONLY CHANGES FORMS, LIKE WHEN CHEMICAL ENERGY IN FOOD BECOMES KINETIC ENERGY WHEN WE MOVE.

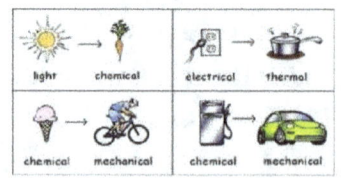

ENERGY TRANSFORMATION
Energy can change form one type to other. You can't create or destroy energy.

2. LIGHT, SOUND, AND ENERGY

LIGHT TRAVELS AT ABOUT 186,282 MILES PER SECOND. IT TAKES JUST OVER EIGHT MINUTES FOR SUNLIGHT TO REACH EARTH.

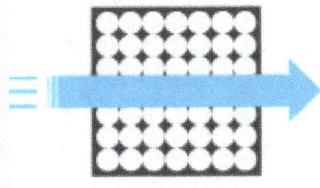

SOUND TRAVELS AT ABOUT 767 MILES PER HOUR. IT MOVES FASTER THROUGH WATER THAN AIR BECAUSE THE MOLECULES IN WATER ARE CLOSER TOGETHER.

KINETIC ENERGY IS THE ENERGY OF MOTION. A MOVING CAR HAS KINETIC ENERGY, AND SO DOES A RUNNING PERSON.

3. EVERYDAY PHYSICS EXAMPLES

BOUNCING A BALL DEMONSTRATES POTENTIAL AND KINETIC ENERGY. THE BALL HAS POTENTIAL ENERGY WHEN HELD UP AND KINETIC ENERGY WHEN IT FALLS.

A SEESAW SHOWS HOW LEVERS WORK. THE PIVOT POINT IN THE MIDDLE HELPS BALANCE THE WEIGHTS ON EITHER END.

ROLLER COASTERS USE THE PRINCIPLES OF GRAVITY AND INERTIA. THE INITIAL CLIMB GIVES THE COASTER POTENTIAL ENERGY, WHICH CONVERTS TO KINETIC ENERGY AS IT SPEEDS DOWN.

EARTH SCIENCE
1. WEATHER, ROCKS, AND MINERALS

CLOUDS ARE MADE OF TINY WATER DROPLETS OR ICE CRYSTALS. THEY FORM WHEN WARM AIR RISES AND COOLS, CAUSING THE WATER VAPOR TO CONDENSE.

DIAMONDS ARE THE HARDEST NATURAL MATERIAL ON EARTH. THEY FORM UNDER HIGH PRESSURE AND TEMPERATURE CONDITIONS DEEP WITHIN THE EARTH.

QUARTZ IS ONE OF THE MOST COMMON MINERALS IN THE EARTH'S CRUST. IT'S USED IN MAKING GLASS, ELECTRONICS, AND WATCHES.

2. VOLCANOES, EARTHQUAKES, AND NATURAL PHENOMENA

VOLCANOES ERUPT WHEN MAGMA FROM BENEATH THE EARTH'S CRUST ESCAPES TO THE SURFACE. THE LARGEST VOLCANO IN OUR SOLAR SYSTEM IS OLYMPUS MONS ON MARS.

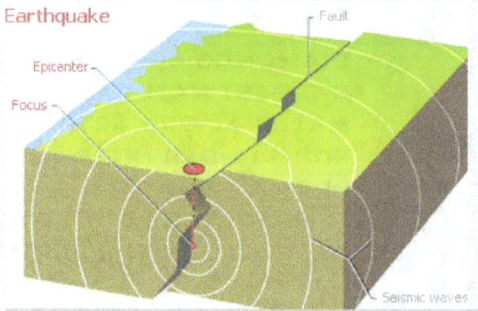

EARTHQUAKES OCCUR WHEN TECTONIC PLATES SHIFT. THE POINT WHERE THE EARTHQUAKE STARTS IS CALLED THE FOCUS.

AURORAS (NORTHERN AND SOUTHERN LIGHTS) ARE NATURAL LIGHT DISPLAYS IN THE EARTH'S SKY. THEY ARE CAUSED BY THE COLLISION OF SOLAR WIND WITH THE EARTH'S MAGNETIC FIELD.

3. THE WATER CYCLE AND CLIMATE CHANGE

THE WATER CYCLE INVOLVES EVAPORATION, CONDENSATION, PRECIPITATION, AND COLLECTION. WATER CONTINUOUSLY MOVES BETWEEN THE EARTH'S SURFACE AND THE ATMOSPHERE.

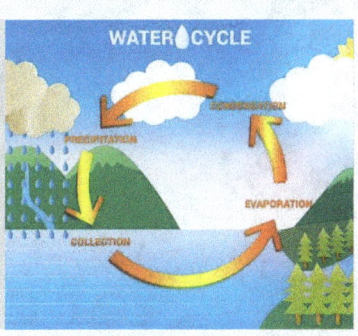

CLIMATE CHANGE IS CAUSED BY FACTORS SUCH AS BURNING FOSSIL FUELS, DEFORESTATION, AND INDUSTRIAL PROCESSES. IT LEADS TO GLOBAL WARMING AND EXTREME WEATHER EVENTS.

GLACIERS STORE ABOUT 69% OF THE WORLD'S FRESH WATER. THEY ARE SLOWLY MELTING DUE TO RISING GLOBAL TEMPERATURES, CONTRIBUTING TO SEA LEVEL RISE.

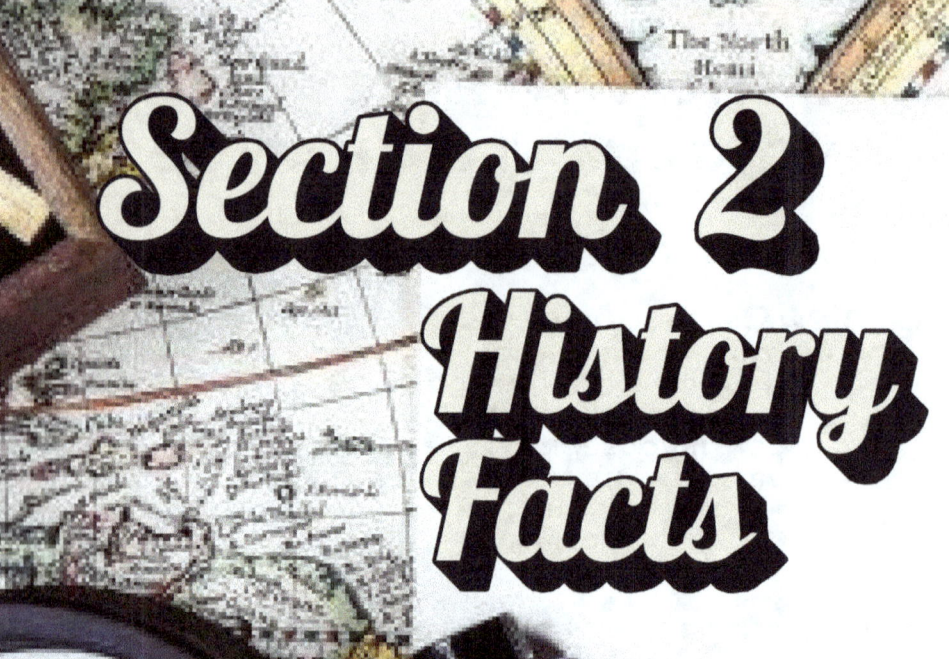

ANCIENT CIVILIZATIONS
1. EGYPTIAN PYRAMIDS AND MUMMIES

THE GREAT PYRAMID OF GIZA IS THE OLDEST OF THE SEVEN WONDERS OF THE ANCIENT WORLD AND THE ONLY ONE STILL IN EXISTENCE.

MUMMIFICATION WAS A PROCESS USED BY ANCIENT EGYPTIANS TO PRESERVE BODIES FOR THE AFTERLIFE. IT INVOLVED REMOVING INTERNAL ORGANS AND WRAPPING THE BODY IN LINEN.

THE SPHINX IS A LIMESTONE STATUE WITH THE BODY OF A LION AND THE HEAD OF A HUMAN. IT STANDS GUARD NEAR THE PYRAMIDS OF GIZA.

2. GREEK MYTHOLOGY AND THE OLYMPICS

ZEUS WAS THE KING OF THE GODS IN GREEK MYTHOLOGY. HE RULED FROM MOUNT OLYMPUS.

THE FIRST OLYMPIC GAMES WERE HELD IN OLYMPIA, GREECE, IN 776 BCE. THEY WERE HELD EVERY FOUR YEARS IN HONOR OF ZEUS.

ATHENA WAS THE GODDESS OF WISDOM AND WAR. THE CITY OF ATHENS IS NAMED AFTER HER.

3. ROMAN EMPIRE AND THEIR ARCHITECTURE

THE COLOSSEUM IN ROME COULD HOLD UP TO 80,000 SPECTATORS. IT WAS USED FOR GLADIATORIAL CONTESTS AND PUBLIC SPECTACLES.

AQUEDUCTS WERE CONSTRUCTED BY THE ROMANS TO TRANSPORT WATER FROM DISTANT SOURCES INTO CITIES AND TOWNS.

JULIUS CAESAR WAS A NOTABLE LEADER OF THE ROMAN EMPIRE. HE PLAYED A CRITICAL ROLE IN THE EVENTS LEADING TO THE DEMISE OF THE ROMAN REPUBLIC AND THE RISE OF THE EMPIRE.

MEDIEVAL TIMES
1. KNIGHTS AND CASTLES

KNIGHTS WERE WARRIORS OF THE MEDIEVAL PERIOD. THEY WORE ARMOR AND FOUGHT ON HORSEBACK.

CASTLES WERE FORTIFIED STRUCTURES BUILT TO PROTECT AGAINST INVADERS. THEY HAD MOATS, DRAWBRIDGES, AND THICK WALLS.

CHIVALRY WAS THE CODE OF CONDUCT FOR KNIGHTS, EMPHASIZING BRAVERY, HONOR, AND RESPECT FOR WOMEN AND THE WEAK.

2. FAMOUS KINGS AND QUEENS

KING ARTHUR IS A LEGENDARY BRITISH LEADER WHO, ACCORDING TO MEDIEVAL HISTORIES AND ROMANCES, LED THE DEFENSE AGAINST SAXON INVADERS.

JOAN OF ARC WAS A FRENCH PEASANT GIRL WHO LED THE FRENCH ARMY TO SEVERAL IMPORTANT VICTORIES DURING THE HUNDRED YEARS' WAR.

CHARLEMAGNE, ALSO KNOWN AS CHARLES THE GREAT, WAS THE KING OF THE FRANKS. HE UNITED MUCH OF WESTERN EUROPE DURING THE EARLY MIDDLE AGES.

3. LIFE IN THE MEDIEVAL PERIOD

FEUDALISM WAS THE DOMINANT SOCIAL SYSTEM. LORDS OWNED LAND AND VASSALS WORKED IT IN EXCHANGE FOR PROTECTION.

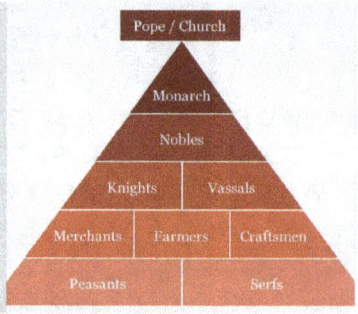

GUILDS WERE ASSOCIATIONS OF ARTISANS OR MERCHANTS WHO CONTROLLED THE PRACTICE OF THEIR CRAFT IN A PARTICULAR TOWN.

MEDIEVAL FAIRS WERE LARGE MARKETS THAT TOOK PLACE IN TOWNS. THEY WERE A MAJOR SOURCE OF ENTERTAINMENT AND TRADE.

MODERN HISTORY
1. IMPORTANT INVENTIONS AND DISCOVERIES

THE PRINTING PRESS, INVENTED BY JOHANNES GUTENBERG IN THE 15TH CENTURY, REVOLUTIONIZED THE WAY INFORMATION WAS SHARED AND MADE BOOKS MORE ACCESSIBLE.

THE STEAM ENGINE, DEVELOPED BY JAMES WATT IN THE 18TH CENTURY, WAS A KEY COMPONENT OF THE INDUSTRIAL REVOLUTION.

PENICILLIN, DISCOVERED BY ALEXANDER FLEMING IN 1928, WAS THE FIRST TRUE ANTIBIOTIC AND HAS SAVED COUNTLESS LIVES.

2. NOTABLE HISTORICAL FIGURES

LEONARDO DA VINCI WAS A RENAISSANCE ARTIST AND INVENTOR. HE PAINTED THE MONA LISA AND DESIGNED NUMEROUS INVENTIONS.

ISAAC NEWTON WAS A MATHEMATICIAN AND PHYSICIST WHO FORMULATED THE LAWS OF MOTION AND UNIVERSAL GRAVITATION.

MAHATMA GANDHI LED INDIA TO INDEPENDENCE FROM BRITISH RULE THROUGH NONVIOLENT RESISTANCE AND CIVIL DISOBEDIENCE.

3. MAJOR EVENTS

THE RENAISSANCE WAS A PERIOD OF GREAT CULTURAL AND ARTISTIC ACHIEVEMENT IN EUROPE, SPANNING THE 14TH TO THE 17TH CENTURY.

THE INDUSTRIAL REVOLUTION BEGAN IN THE LATE 18TH CENTURY AND TRANSFORMED MANUFACTURING PROCESSES WITH MACHINES AND FACTORIES.

WORLD WAR II, FOUGHT FROM 1939 TO 1945, INVOLVED MANY OF THE WORLD'S NATIONS AND HAD A PROFOUND IMPACT ON GLOBAL HISTORY.

INDIAN HISTORY
1. ANCIENT INDIA

INDUS VALLEY CIVILIZATION IS ONE OF THE WORLD'S OLDEST CIVILIZATIONS, KNOWN FOR ITS ADVANCED URBAN PLANNING AND ARCHITECTURE.

ASHOKA THE GREAT WAS A MAURYAN EMPEROR WHO SPREAD BUDDHISM AND ESTABLISHED A VAST EMPIRE IN INDIA.

ARYABHATA, AN ANCIENT INDIAN MATHEMATICIAN AND ASTRONOMER, MADE SIGNIFICANT CONTRIBUTIONS TO MATHEMATICS AND ASTRONOMY.

2. MEDIEVAL INDIA

THE GUPTA EMPIRE IS KNOWN AS THE GOLDEN AGE OF INDIA DUE TO ITS ADVANCEMENTS IN SCIENCE, MATHEMATICS, AND ART.

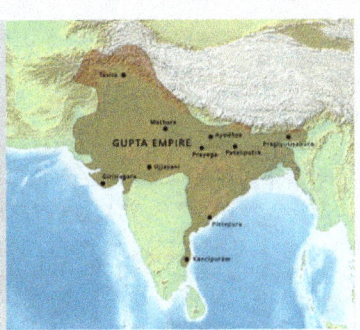

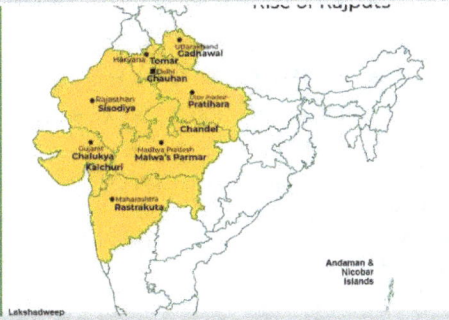

RAJPUT KINGDOMS WERE WARRIOR CLANS KNOWN FOR THEIR VALOR AND CHIVALRY. THEY BUILT MAGNIFICENT FORTS AND PALACES.

THE DELHI SULTANATE WAS A PERIOD OF MUSLIM RULE IN INDIA, KNOWN FOR ITS CULTURAL AND ARCHITECTURAL ACHIEVEMENTS.

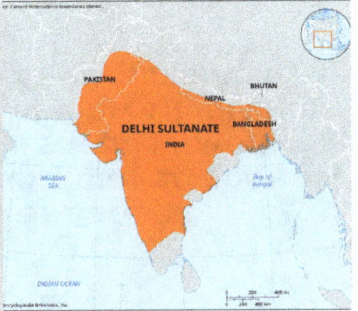

3. MUGHAL INDIA

AKBAR THE GREAT WAS A MUGHAL EMPEROR WHO EXPANDED THE EMPIRE AND PROMOTED CULTURAL AND RELIGIOUS TOLERANCE.

TAJ MAHAL, BUILT BY EMPEROR SHAH JAHAN, IS A UNESCO WORLD HERITAGE SITE AND A SYMBOL OF INDIA'S RICH HISTORY.

AURANGZEB WAS THE LAST SIGNIFICANT MUGHAL RULER, KNOWN FOR HIS MILITARY CONQUESTS AND EXPANSION OF THE EMPIRE.

4. MODERN INDIA

INDIAN INDEPENDENCE MOVEMENT WAS LED BY FIGURES LIKE MAHATMA GANDHI, JAWAHARLAL NEHRU, AND SUBHAS CHANDRA BOSE.

INDEPENDENCE FROM BRITISH RULE WAS ACHIEVED ON AUGUST 15, 1947.

DR. B.R. AMBEDKAR WAS THE PRINCIPAL ARCHITECT OF THE INDIAN CONSTITUTION AND A CHAMPION FOR SOCIAL JUSTICE.

HUMAN BODY
1. AMAZING FACTS ABOUT OUR ORGANS

THE BRAIN HAS ABOUT 86 BILLION NEURONS. IT'S THE COMMAND CENTER OF THE BODY.

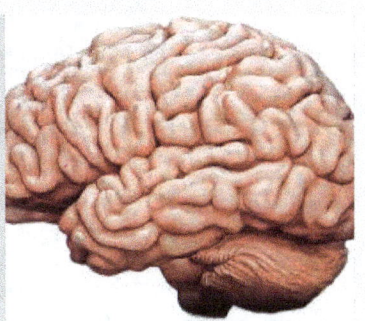

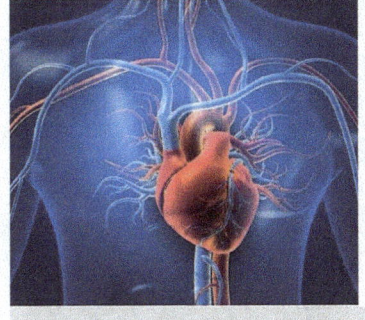

THE HEART PUMPS APPROXIMATELY 2,000 GALLONS OF BLOOD EACH DAY.

THE LIVER IS THE LARGEST INTERNAL ORGAN AND HAS OVER 500 FUNCTIONS, INCLUDING DETOXIFYING HARMFUL SUBSTANCES.

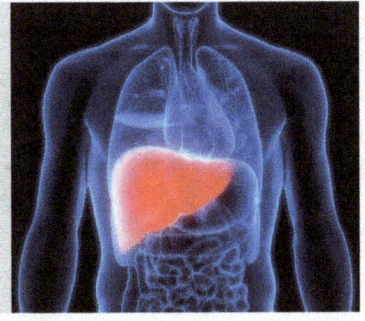

2. OUR SENSES

EYES CAN DISTINGUISH ABOUT 10 MILLION DIFFERENT COLORS. THE HUMAN EYE CAN DETECT A CANDLE FLAME FROM 1.7 MILES AWAY IN THE DARK.

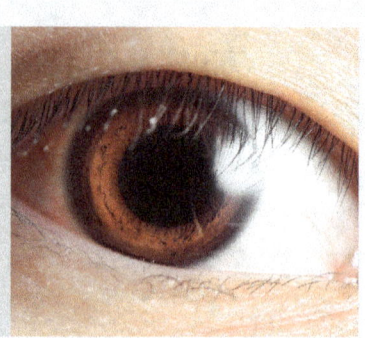

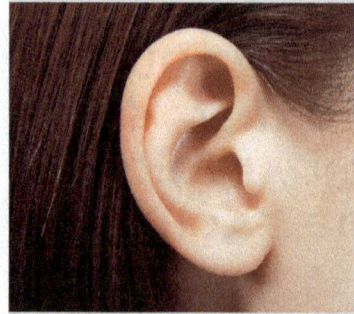

EARS NOT ONLY HELP US HEAR BUT ALSO PLAY A KEY ROLE IN MAINTAINING BALANCE.

TASTE BUDS ON OUR TONGUE CAN DETECT SWEET, SOUR, SALTY, BITTER, AND UMAMI FLAVORS.

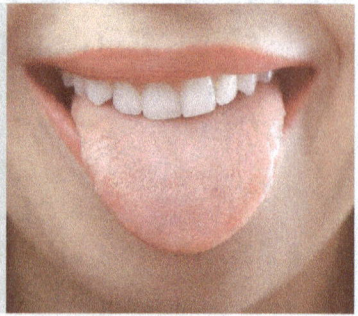

3. FUN HUMAN CAPABILITIES

FINGERNAILS GROW FASTER ON THE HAND YOU WRITE WITH. THEY ALSO GROW MORE QUICKLY IN SUMMER THAN IN WINTER.

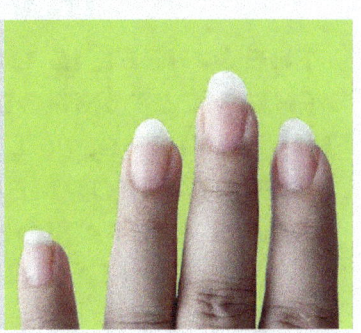

HUMANS SHARE 99% OF THEIR DNA WITH CHIMPANZEES, OUR CLOSEST LIVING RELATIVES.

LAUGHING 100 TIMES IS EQUIVALENT TO 15 MINUTES OF EXERCISE ON A STATIONARY BIKE IN TERMS OF CALORIES BURNED.

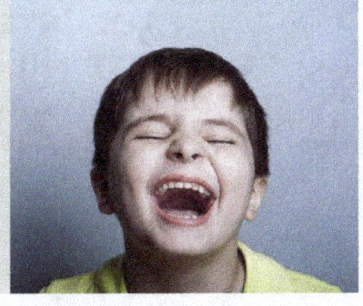

INVENTIONS AND DISCOVERIES
1. GROUNDBREAKING INVENTIONS

THE WHEEL IS ONE OF THE MOST IMPORTANT INVENTIONS. IT LED TO THE DEVELOPMENT OF TRANSPORTATION AND MACHINERY.

ELECTRICITY REVOLUTIONIZED THE WORLD. THOMAS EDISON'S INVENTION OF THE ELECTRIC LIGHT BULB CHANGED HOW WE LIVE.

THE INTERNET CONNECTS MILLIONS OF PEOPLE AROUND THE WORLD. IT WAS DEVELOPED BY ARPANET IN THE LATE 20TH CENTURY.

2. FAMOUS INVENTORS

ALEXANDER GRAHAM BELL INVENTED THE TELEPHONE. HIS WORK MADE INSTANT COMMUNICATION OVER LONG DISTANCES POSSIBLE.

THE WRIGHT BROTHERS, ORVILLE AND WILBUR, INVENTED THE FIRST SUCCESSFUL AIRPLANE, LEADING TO THE ERA OF MODERN AVIATION.

MARIE CURIE WAS THE FIRST WOMAN TO WIN A NOBEL PRIZE AND DISCOVERED RADIUM AND POLONIUM.

3. SURPRISING DISCOVERIES

GRAVITY WAS DISCOVERED BY SIR ISAAC NEWTON WHEN HE SAW AN APPLE FALL FROM A TREE.

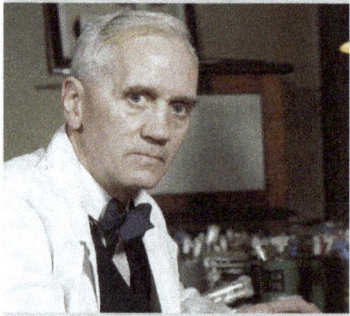

PENICILLIN WAS DISCOVERED BY ALEXANDER FLEMING FROM MOLD. IT BECAME THE WORLD'S FIRST ANTIBIOTIC.

THE ROSETTA STONE WAS DISCOVERED IN EGYPT AND HELPED DECODE EGYPTIAN HIEROGLYPHS.

WORLD RECORDS
1. TALLEST, FASTEST, SMALLEST, AND LARGEST RECORDS

THE TALLEST BUILDING IN THE WORLD IS THE BURJ KHALIFA IN DUBAI, STANDING AT 828 METERS.

THE FASTEST ANIMAL ON LAND IS THE CHEETAH, WHICH CAN RUN UP TO 75 MPH.

THE SMALLEST COUNTRY IN THE WORLD IS VATICAN CITY, WITH AN AREA OF JUST 0.2 SQUARE MILES.

2. EXTRAORDINARY HUMAN FEATS

MICHAEL PHELPS HOLDS THE RECORD FOR THE MOST OLYMPIC GOLD MEDALS, WITH A TOTAL OF 23.

USAIN BOLT IS THE FASTEST SPRINTER, HOLDING THE WORLD RECORD FOR THE 100 METERS AT 9.58 SECONDS.

JEANNE CALMENT LIVED TO BE 122 YEARS OLD, MAKING HER THE OLDEST RECORDED HUMAN.

3. AMAZING RECORDS IN NATURE AND TECHNOLOGY

MOUNT EVEREST IS THE HIGHEST MOUNTAIN, REACHING 29,029 FEET ABOVE SEA LEVEL.

THE MARIANA TRENCH IS THE DEEPEST PART OF THE WORLD'S OCEANS, PLUNGING ABOUT 36,000 FEET.

THE INTERNATIONAL SPACE STATION IS THE LARGEST HUMAN-MADE STRUCTURE IN ORBIT, SPANNING THE SIZE OF A FOOTBALL FIELD.

LANGUAGE AND LITERATURE
1. INTERESTING FACTS ABOUT LANGUAGES

MANDARIN CHINESE IS THE MOST SPOKEN LANGUAGE IN THE WORLD, WITH OVER A BILLION SPEAKERS.

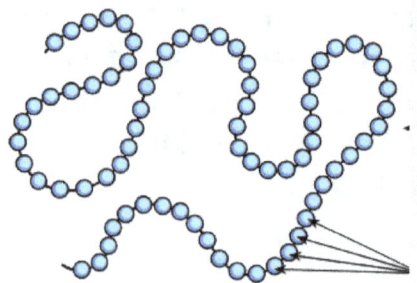

THE LONGEST WORD IN THE ENGLISH LANGUAGE IS 189,819 LETTERS LONG, RELATED TO A TYPE OF PROTEIN.

ESPERANTO IS A CONSTRUCTED LANGUAGE CREATED TO FOSTER INTERNATIONAL COMMUNICATION.

2. FAMOUS AUTHORS AND THEIR WORKS

WILLIAM SHAKESPEARE WROTE CLASSICS LIKE "ROMEO AND JULIET" AND "HAMLET." HE IS CONSIDERED ONE OF THE GREATEST WRITERS IN THE ENGLISH LANGUAGE.

J.K. ROWLING CREATED THE HARRY POTTER SERIES, WHICH HAS CAPTIVATED READERS AROUND THE GLOBE.

MARK TWAIN, AUTHOR OF "THE ADVENTURES OF TOM SAWYER" AND "ADVENTURES OF HUCKLEBERRY FINN," IS KNOWN FOR HIS WIT AND SOCIAL CRITICISM.

3. UNIQUE LITERARY TRADITIONS

HAIKU IS A TRADITIONAL FORM OF JAPANESE POETRY WITH THREE LINES AND A 5-7-5 SYLLABLE STRUCTURE.

heavy grey, pure white
a drama of changing sky
blinding silver, blue

EPIC POEMS LIKE "THE ILIAD" AND "THE ODYSSEY" BY HOMER TELL GRAND TALES OF HEROISM AND ADVENTURE.

FABLES ARE SHORT STORIES THAT TEACH MORAL LESSONS, OFTEN FEATURING ANIMALS WITH HUMAN TRAITS.

CONTINENTS AND COUNTRIES

1. FACTS ABOUT EACH CONTINENT

ASIA IS THE LARGEST CONTINENT, BOTH IN SIZE AND POPULATION. IT IS HOME TO THE HIGHEST POINT ON EARTH, MOUNT EVEREST.

 AFRICA HAS THE LONGEST RIVER, THE NILE, AND THE LARGEST DESERT, THE SAHARA

NORTH AMERICA INCLUDES THE THIRD-LARGEST COUNTRY BY AREA, THE UNITED STATES, AND THE LONGEST COASTLINE.

2. UNIQUE ASPECTS OF DIFFERENT COUNTRIES

INDIA HAS THE LARGEST DEMOCRACY IN THE WORLD AND IS FAMOUS FOR ITS DIVERSE CULTURE AND HISTORY.

BRAZIL IS KNOWN FOR THE AMAZON RAINFOREST AND THE ANNUAL CARNIVAL FESTIVAL.

AUSTRALIA IS THE ONLY COUNTRY THAT IS ALSO A CONTINENT AND IS FAMOUS FOR ITS UNIQUE WILDLIFE, LIKE KANGAROOS AND KOALAS.

3. CULTURAL DIVERSITY AROUND THE WORLD

JAPAN IS KNOWN FOR ITS TRADITIONAL TEA CEREMONIES AND VIBRANT FESTIVALS LIKE HANAMI, THE CHERRY BLOSSOM FESTIVAL.

MEXICO CELEBRATES DIA DE LOS MUERTOS (DAY OF THE DEAD), A COLORFUL HOLIDAY HONORING DECEASED LOVED ONES.

ITALY IS FAMOUS FOR ITS RICH HISTORY, ART, AND CUISINE, INCLUDING PASTA AND PIZZA.

MOUNTAINS, RIVERS, AND DESERTS

1. TALLEST MOUNTAINS AND LONGEST RIVERS

MOUNT EVEREST IN THE HIMALAYAS IS THE TALLEST MOUNTAIN IN THE WORLD, STANDING AT 29,029 FEET.

THE NILE RIVER IN AFRICA IS THE LONGEST RIVER, STRETCHING ABOUT 4,135 MILES.

MOUNT KILIMANJARO IN TANZANIA IS THE HIGHEST FREE-STANDING MOUNTAIN IN THE WORLD.

2. LARGEST DESERTS AND THEIR FEATURES

THE SAHARA DESERT IN AFRICA IS THE LARGEST HOT DESERT, COVERING 9 MILLION SQUARE KILOMETERS.

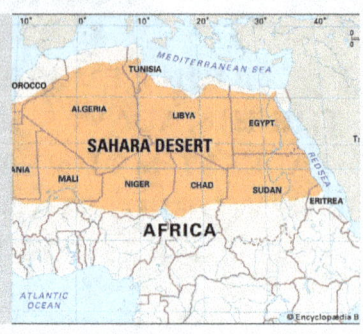

THE ANTARCTIC DESERT IS THE COLDEST AND DRIEST DESERT, COVERING THE ENTIRE CONTINENT OF ANTARCTICA.

THE GOBI DESERT IN ASIA IS KNOWN FOR ITS HARSH CLIMATE AND UNIQUE LANDSCAPES.

3. MAJOR NATURAL LANDMARKS

THE GRAND CANYON IN THE UNITED STATES IS A MASSIVE GORGE CARVED BY THE COLORADO RIVER.

VICTORIA FALLS ON THE BORDER OF ZAMBIA AND ZIMBABWE IS ONE OF THE LARGEST AND MOST FAMOUS WATERFALLS IN THE WORLD.

THE GREAT BARRIER REEF IN AUSTRALIA IS THE WORLD'S LARGEST CORAL REEF SYSTEM, STRETCHING OVER 1,400 MILES.

OCEANS AND SEAS
1. INTERESTING MARINE LIFE

BLUE WHALES ARE THE LARGEST ANIMALS ON EARTH AND CAN WEIGH AS MUCH AS 200 TONS.

OCTOPUSES ARE INCREDIBLY INTELLIGENT AND CAN CHANGE COLOR AND TEXTURE TO BLEND IN WITH THEIR SURROUNDINGS.

CORAL REEFS ARE MADE UP OF TINY ANIMALS CALLED POLYPS AND ARE HOME TO A DIVERSE RANGE OF MARINE SPECIES.

2. MAJOR OCEANS AND SEAS

THE PACIFIC OCEAN IS THE LARGEST AND DEEPEST OCEAN, COVERING MORE THAN 63 MILLION SQUARE MILES.

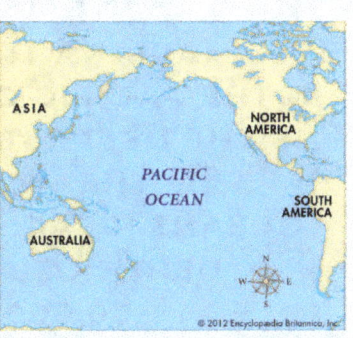

THE ATLANTIC OCEAN IS KNOWN FOR ITS SARGASSO SEA, WHICH HAS NO SHORES AND IS SURROUNDED BY OCEAN CURRENTS.

THE INDIAN OCEAN IS THE WARMEST OCEAN AND IS FAMOUS FOR ITS MONSOON WEATHER PATTERNS.

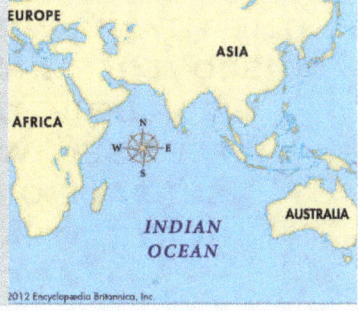

3. FACTS ABOUT UNDERWATER EXPLORATION

THE MARIANA TRENCH IN THE PACIFIC OCEAN IS THE DEEPEST PART OF THE WORLD'S OCEANS, REACHING A DEPTH OF ABOUT 36,000 FEET.

SUBMARINES AND ROVS (REMOTELY OPERATED VEHICLES) ARE USED TO EXPLORE THE DEEP SEA, WHERE SUNLIGHT CANNOT REACH.

THE TITANIC WRECK, DISCOVERED IN 1985, LIES ABOUT 12,500 FEET BELOW THE SURFACE OF THE NORTH ATLANTIC OCEAN.

MAPS AND GLOBES
1. HOW TO READ MAPS AND GLOBES

MAPS SHOW THE EARTH'S SURFACE ON A FLAT PIECE OF PAPER AND INCLUDE SYMBOLS, LEGENDS, AND SCALES TO REPRESENT DIFFERENT FEATURES.

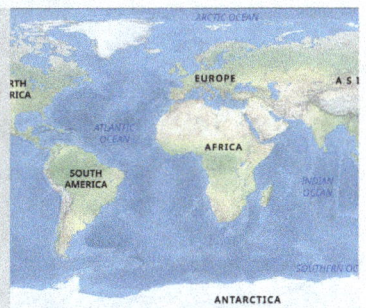

GLOBES ARE THREE-DIMENSIONAL MODELS OF THE EARTH, SHOWING CONTINENTS, OCEANS, AND MAJOR GEOGRAPHICAL FEATURES.

LATITUDE AND LONGITUDE LINES HELP PINPOINT EXACT LOCATIONS ON THE EARTH'S SURFACE.

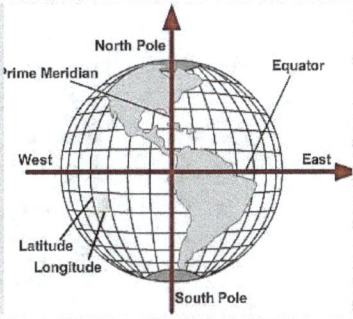

2. IMPORTANT LINES LIKE THE EQUATOR AND PRIME MERIDIAN

THE EQUATOR IS AN IMAGINARY LINE THAT DIVIDES THE EARTH INTO THE NORTHERN AND SOUTHERN HEMISPHERES.

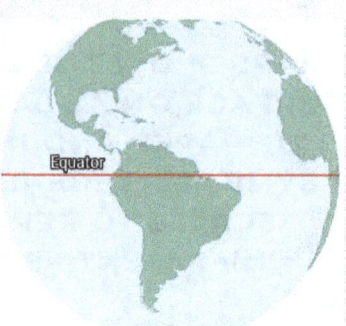

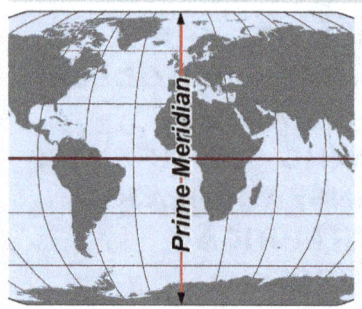

THE PRIME MERIDIAN IS AN IMAGINARY LINE THAT DIVIDES THE EARTH INTO THE EASTERN AND WESTERN HEMISPHERES. IT RUNS THROUGH GREENWICH, ENGLAND.

THE TROPIC OF CANCER AND THE TROPIC OF CAPRICORN MARK THE BOUNDARIES OF THE TROPICS, WHERE THE SUN CAN BE DIRECTLY OVERHEAD.

3. FASCINATING CARTOGRAPHY FACTS

CARTOGRAPHY IS THE SCIENCE OF MAKING MAPS. EARLY MAPS WERE OFTEN HAND-DRAWN AND INCLUDED MYTHICAL CREATURES AND INACCURATE FEATURES.

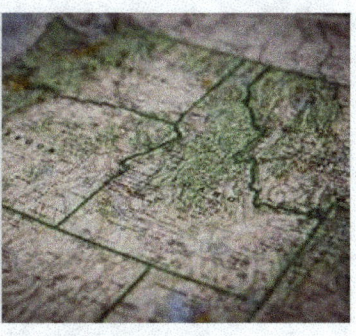

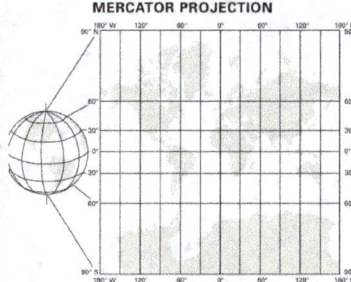

THE MERCATOR PROJECTION IS A FAMOUS MAP PROJECTION THAT DISTORTS SIZES BUT PRESERVES ANGLES, MAKING IT USEFUL FOR NAVIGATION.

TOPOGRAPHIC MAPS SHOW ELEVATION AND TERRAIN FEATURES, HELPING HIKERS AND GEOLOGISTS UNDERSTAND THE LANDSCAPE.

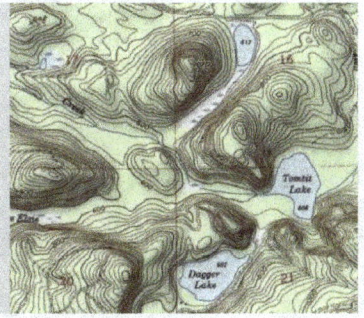

Section 5
Space Facts

SOLAR SYSTEM
1. PLANETS, MOONS, AND THE SUN

MERCURY IS THE SMALLEST PLANET AND CLOSEST TO THE SUN, WITH EXTREME TEMPERATURES.

VENUS IS THE HOTTEST PLANET, WITH SURFACE TEMPERATURES REACHING UP TO 900 DEGREES FAHRENHEIT.

EARTH IS THE ONLY PLANET KNOWN TO SUPPORT LIFE. IT HAS ONE MOON THAT AFFECTS TIDES.

2. INTERESTING FACTS ABOUT EACH PLANET

MARS IS KNOWN AS THE RED PLANET BECAUSE OF ITS IRON OXIDE (RUST) SURFACE.

JUPITER IS THE LARGEST PLANET IN OUR SOLAR SYSTEM AND HAS A GIANT STORM CALLED THE GREAT RED SPOT.

SATURN IS FAMOUS FOR ITS STUNNING RING SYSTEM, MADE OF ICE AND ROCK PARTICLES.

3. ASTEROIDS, COMETS, AND METEOROIDS

ASTEROIDS ARE ROCKY OBJECTS ORBITING THE SUN, MOSTLY FOUND IN THE ASTEROID BELT BETWEEN MARS AND JUPITER.

COMETS ARE ICY BODIES THAT DEVELOP TAILS WHEN THEY APPROACH THE SUN, LIKE HALLEY'S COMET, WHICH APPEARS EVERY 76 YEARS.

METEOROIDS ARE SMALL ROCKY OR METALLIC BODIES IN SPACE. WHEN THEY ENTER EARTH'S ATMOSPHERE AND BURN UP, THEY ARE CALLED METEORS OR SHOOTING STARS.

STARS AND GALAXIES
1. DIFFERENT TYPES OF STARS

DWARF STARS ARE SMALL STARS, LIKE OUR SUN, WHICH IS A YELLOW DWARF.

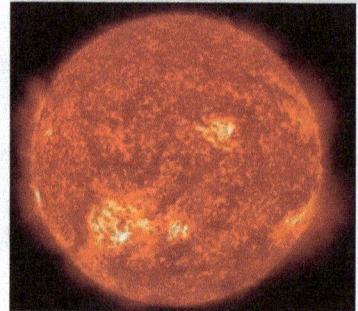

GIANT STARS ARE MUCH LARGER AND BRIGHTER THAN DWARF STARS. RED GIANTS ARE STARS IN THE LATE STAGES OF THEIR LIFE.

NEUTRON STARS ARE INCREDIBLY DENSE REMNANTS OF SUPERNOVA EXPLOSIONS, WITH INTENSE GRAVITATIONAL FIELDS.

2. FACTS ABOUT THE MILKY WAY AND OTHER GALAXIES

THE MILKY WAY IS OUR HOME GALAXY, CONTAINING ABOUT 100 BILLION STARS.

ANDROMEDA IS THE CLOSEST SPIRAL GALAXY TO THE MILKY WAY AND IS ON A COLLISION COURSE WITH IT IN ABOUT 4 BILLION YEARS.

ELLIPTICAL GALAXIES ARE SHAPED LIKE ELONGATED SPHERES AND CONTAIN OLDER, REDDER STARS.

3. THE LIFE CYCLE OF STARS

NEBULAS ARE CLOUDS OF GAS AND DUST WHERE STARS ARE BORN.

A STAR'S MAIN SEQUENCE PHASE IS WHEN IT FUSES HYDROGEN INTO HELIUM, PRODUCING LIGHT AND HEAT.

SUPERNOVAE ARE EXPLOSIVE DEATHS OF MASSIVE STARS, OFTEN LEAVING BEHIND NEUTRON STARS OR BLACK HOLES.

SPACE EXPLORATION
1. FAMOUS SPACE MISSIONS AND ASTRONAUTS

APOLLO 11 WAS THE FIRST MANNED MISSION TO LAND ON THE MOON IN 1969, WITH NEIL ARMSTRONG AND BUZZ ALDRIN.

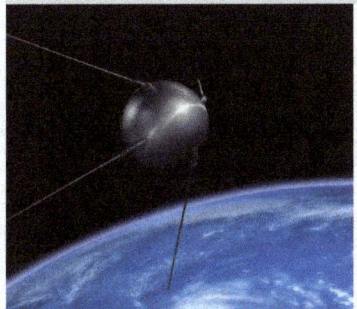

SPUTNIK 1 WAS THE FIRST ARTIFICIAL SATELLITE, LAUNCHED BY THE SOVIET UNION IN 1957.

VALENTINA TERESHKOVA WAS THE FIRST WOMAN IN SPACE, ORBITING EARTH IN 1963 ABOARD VOSTOK 6.

2. INTERESTING FACTS ABOUT SPACE TRAVEL

THE INTERNATIONAL SPACE STATION (ISS) ORBITS EARTH AND SERVES AS A SPACE LABORATORY FOR INTERNATIONAL ASTRONAUTS.

SPACE SUITS PROTECT ASTRONAUTS FROM EXTREME TEMPERATURES AND LACK OF OXYGEN IN SPACE.

MICROGRAVITY IN SPACE CAUSES MUSCLES AND BONES TO WEAKEN, SO ASTRONAUTS EXERCISE DAILY TO STAY HEALTHY.

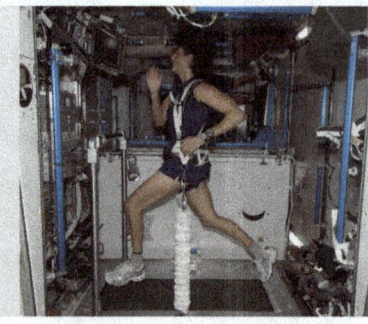

3. THE INTERNATIONAL SPACE STATION AND FUTURE MISSIONS

THE ISS TRAVELS AT ABOUT 17,500 MILES PER HOUR AND ORBITS EARTH APPROXIMATELY EVERY 90 MINUTES.

MARS MISSIONS AIM TO EXPLORE THE RED PLANET FOR SIGNS OF PAST OR PRESENT LIFE AND POTENTIAL FUTURE HUMAN HABITATION.

ARTEMIS PROGRAM PLANS TO RETURN HUMANS TO THE MOON AND ESTABLISH A SUSTAINABLE PRESENCE BY THE LATE 2020S.

UNIVERSE MYSTERIES
1. BLACK HOLES AND DARK MATTER

BLACK HOLES ARE REGIONS OF SPACE WHERE GRAVITY IS SO STRONG THAT NOT EVEN LIGHT CAN ESCAPE. THEY ARE FORMED FROM THE REMNANTS OF MASSIVE STARS.

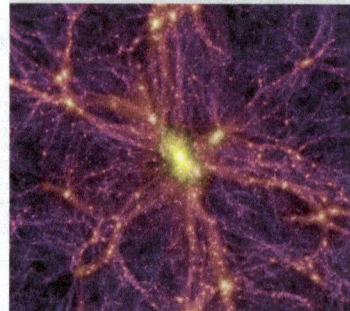

DARK MATTER MAKES UP ABOUT 27% OF THE UNIVERSE. IT'S INVISIBLE AND DOESN'T EMIT LIGHT OR ENERGY, BUT ITS PRESENCE CAN BE INFERRED FROM GRAVITATIONAL EFFECTS.

EVENT HORIZON IS THE BOUNDARY AROUND A BLACK HOLE BEYOND WHICH NOTHING CAN RETURN.

2. THEORIES ABOUT THE ORIGIN OF THE UNIVERSE

THE BIG BANG THEORY SUGGESTS THE UNIVERSE BEGAN ABOUT 13.8 BILLION YEARS AGO FROM A SINGLE POINT OF EXTREMELY HIGH DENSITY AND TEMPERATURE.

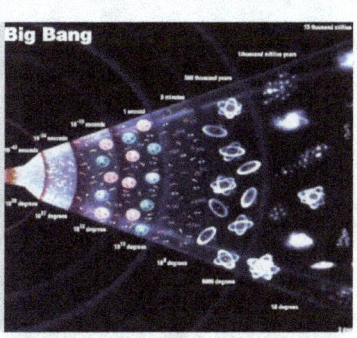

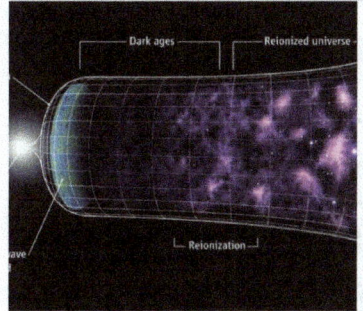

COSMIC INFLATION PROPOSES A RAPID EXPANSION OF THE UNIVERSE RIGHT AFTER THE BIG BANG.

MULTIVERSE THEORY SPECULATES THAT OUR UNIVERSE IS JUST ONE OF MANY POSSIBLE UNIVERSES WITH DIFFERENT PHYSICAL LAWS.

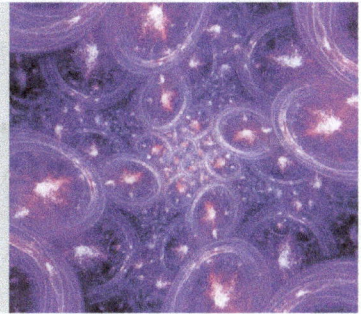

3. FUN TRIVIA ABOUT SPACE PHENOMENA

PULSARS ARE ROTATING NEUTRON STARS THAT EMIT BEAMS OF RADIATION. THEY APPEAR TO PULSE BECAUSE THEIR MAGNETIC POLES POINT TOWARD EARTH AT REGULAR INTERVALS.

QUASARS ARE EXTREMELY BRIGHT AND DISTANT OBJECTS POWERED BY SUPERMASSIVE BLACK HOLES AT THEIR CENTERS.

COSMIC MICROWAVE BACKGROUND IS THE AFTERGLOW OF THE BIG BANG, PROVIDING A SNAPSHOT OF THE EARLY UNIVERSE.

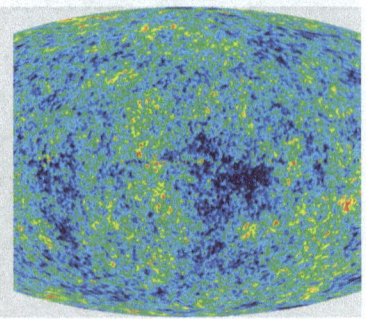

Test your Knowledge

1. What is the largest land animal?
A) Elephant
B) Blue whale
C) Giraffe
D) Rhino

2. Which planet is known as the Red Planet?
A) Venus
B) Mars
C) Jupiter
D) Saturn

3. Who invented the telephone?
A) Thomas Edison
B) Alexander Graham Bell
C) Nikola Tesla
D) James Watt

4. What is the fastest land animal?
A) Cheetah
B) Lion
C) Gazelle
D) Tiger

5. What is the smallest country in the world?
A) Monaco
B) Liechtenstein
C) Vatican City
D) San Marino

6. Which organ is the largest in the human body?
A) Brain
B) Liver
C) Skin
D) Heart

7. Which ancient civilization is known for building pyramids?
A) Romans
B) Greeks
C) Egyptians
D) Mayans

8. What is the longest river in the world?
A) Amazon River
B) Nile River
C) Yangtze River
D) Mississippi River

9. Who was the first woman in space?
A) Sally Ride
B) Valentina Tereshkova
C) Mae Jemison
D) Kalpana Chawla

10. What does a Venus Flytrap eat?
A) Water
B) Soil
C) Insects
D) Leaves

11. Which country is famous for the Great Barrier Reef?
A) Brazil
B) Australia
C) India
D) South Africa

12. What is the name of the galaxy we live in?
A) Andromeda
B) Milky Way
C) Triangulum
D) Sombrero

13. Which element is the most abundant in the Earth's crust?
A) Hydrogen
B) Oxygen
C) Iron
D) Silicon

14. Who wrote "Romeo and Juliet"?
A) Charles Dickens
B) William Shakespeare
C) Mark Twain
D) J.K. Rowling

15. What is the process by which plants make their food using sunlight?
A) Respiration
B) Photosynthesis
C) Fermentation
D) Digestion

16. Which planet is known for its stunning rings?
A) Jupiter
B) Uranus
C) Saturn
D) Neptune

17. Who was the principal architect of the Indian Constitution?
A) Jawaharlal Nehru
B) Mahatma Gandhi
C) Dr. B.R. Ambedkar
D) Sardar Patel

18. What is the hardest natural material on Earth?
A) Quartz
B) Diamond
C) Gold
D) Steel

19. What causes the Northern and Southern Lights?
A) Meteor showers
B) Solar wind
C) Volcanic eruptions
D) Ocean tides

20. What is the deepest part of the world's oceans?
A) Mariana Trench
B) Bermuda Triangle
C) Java Trench
D) Puerto Rico Trench

Answer Key

1. (A) Elephant
2. (B) Mars
3. (B) Alexander Graham Bell
4. (A) Cheetah
5. (C) Vatican City
6. (C) Skin
7. (C) Egyptians
8. (B) Nile River
9. (B) Valentina Tereshkova
10. (C) Insects

11. (B) Australia
12. (B) Milky Way
13. (B) Oxygen
14. (B) William Shakespeare
15. (B) Photosynthesis
16. (C) Saturn
17. (C) Dr. B.R. Ambedkar
18. (B) Diamond
19. (B) Solar wind
20. (A) Mariana Trench

Copyright © Vansh Tiwari
All rights reserved.

Image credits: Internet

www.ingramcontent.com/pod-product-compliance
Lightning Source LLC
Chambersburg PA
CBHW071951210526
45479CB00003B/890